SOME USEFUL PLANTS OF EARLY NEW ENGLAND

by

JANE STRICKLAND HUSSEY

THE CHANNINGS
Marion Massachusetts

Library of Congress Catalog Card Number 76-23347
ISBN 0-9600496-5-7

Printed in the United States of America

PREFACE

Both the layman and serious botanist will appreciate this gathering together of information which appeared in a recent issue of *Economic Botany*. Many of the plants were here in New England when the early voyagers explored along the New England coast. They have spread all over the United States and are well known to plant lovers. Discussed are aromatic, culinary and medicinal herbs, dye plants, berries, ground covers, shrubs, and trees. They were well known to the Indians, and many have multiple uses.

Jane Strickland Hussey is the Horticulturist at the Plimoth Plantation and has been involved with their gardens and plantings since they opened to the public in 1958. A graduate of the Amy Sacker School of Design in Boston, she attended, one summer, a course in Decorative Arts which was held in the beautiful palace at Fontainbleau, France. Her brother, Charles R. Strickland, first interested her in herbs, and from there she went on to serious research at the Massachusetts Horticultural Society Library, the Boston Athenaeum, and Dumbarton Oaks Library.

The accurate and charming illustrations have been done by Catherine Russell Hammond and Priscilla Heath Kunhardt.

Cover by Catherine Elliott Hedge.

CONTENTS

INTRODUCTION

Our first knowledge of native plants of New England comes from the early voyages along this coast, some long before the first permanent settlement of 1620. John Brereton's *Discovery of the North Part of Virginia*, 1602, and James Rosier's *The Land of Virginia*, 1605, include this area. Other accounts are *Champlain's Voyages, 1604-16*; Christopher Lovett's *A Voyage into New England, 1623-24*; and *Sailors' Narratives Along the New England Coast, 1524-1624*. A sampling of quotations from these voyages follows:

> "Passing up the river we saw certain cottages abandoned by the Savages, and not farre off we behold their Gardens . . . was sown Tobacco, pompions, cucumbers, and such like . . . maize or Indian Wheate among the rest. In the fields we found Wild Pease, Strawberries, very faire and bigge, gooseberries, Raspices, Hurts, and other wild fruits."

> "Before we reached their cabins, we entered a field planted with Indian corn . . . We saw many Brazilian beans, and many squashes of various sizes, very good for eating, some tobacco and roots which they cultivate, the latter having the taste of Artichoke." (Jerusalem Artichoke)

> "The eight and twentieth we entered counsel about abode and plantation, which was concluded to be in the West Part of Elizabeth's Island. It is over blown with wood and rubbish, *viz.* oaks, ashes, beech, walnut, witch hazel, sassafras, and cedars, with divers others of unknown names. The rubbish is wild pease, young sassafras, cherry trees, vines, eglantine, gooseberry bushes, hawthorn, honeysuckles, with others of like quality. The herbs and roots are strawberries, raspberries, ground nuts, alexander, tansy, etc., without count."

There are some mystery plants. I have the feeling that the English, discovering these shores for the first time, gave names to plant life found here, which resembled ones familiar to them at home. "And divers other wholesome herbs, both for profit and pleasure, with great store of aniseed." "Balme, divers other good herbs are there, that grow without industry of man." Professor Gray of Harvard thought that this Balme was a species of Monarda. Another plant which I am ignorant of is Carvell. "Also divers excellent pot herbs grow abundantly among the grasse . . . carvel." Still another is Dragon. John Josselyn lists it as a native herb and wrote "Dragons, their leaves differ from all the known kinds with us, they come up in June." "Dragon-Root (Arisaema dracontium) is a tall tuberous herb with greenish flowers like bracts within which are produced orange red berries. It is related to Jack-in-the-Pulpit." One of the common names of Skunk Cabbage *(Symplocarpus foetidus)* is Dracontium. In England Dragon (Dracontium) is a plant whose lower stalk resembles a snake. John Josselyn speaks of "Muschater, as well as in England." This is the only reference I have ever found of this plant. Of Satyrion, John Josselyn notes "Dogstones, a kind of Satyrion, whereof there are several kinds groweth in our Salt Marshes." The only other reference I have found of Satyrion is in *A New Herbal*, 1619, by Dodoens who listed "right Satyrion or Dioscorides satyrion." There are innumerable quotes of "Wild Pease" throughout all the early literature. Possibly a vetch?

Jane Strickland Hussey

SOME USEFUL PLANTS
OF EARLY NEW ENGLAND

Plants

ADDER'S TONGUE

Ophioglossum vulgatum Ophioglossaceae

A fern which is native to England and New England. "Adder's Tongue comes not up till June; I have found it upon dry hilly grounds, in places where the water hath stood all Winter, in August, and did then make Oyntment of the Herb new gathered; the fairest leaves grow amongst short Hawthorn Bushes, that are plentifully growing in such hollow places." – John Josselyn. Best known as a wound herb, Gerard, in his *Grete Herbal* recommends that, boiled with oil it was "most excellent for green wounds." "It also Stayeth green and fresh wounds. It also stayeth and refresheth all inflammations that arise upon pains by hurts and wounds." – Culpeper.

Adder's Tongue

ALEXANDER

Zizia aurea Umbelliferae

Alexander was noted by John Josselyn as follows: "Alexanders, which grow upon rocks by the Seashore." Thomas Morton adds: "The country there naturally affordeth very goode pot herbes and sallet herbes and those of a more masculine vertue than any of the same species in England . . . Alexanders." Two early voyagers, Gosnold and Bartholomew, in 1602, wrote: "In the mean we sustained ourselves with Alexanders." This pot and salad herb, known also as Early or Golden Parsnip, is a bog or marsh plant.

ARROWHEAD

Sagittaria latifolia Alismataceae

Arrowhead, an aquatic plant, has a leaf which resembles the head of a barbed arrow and is native both to England and New England. Other common names are Smallwater Archer, Adder's Tongue, Moses-in-the-Bulrushes and Suck Potatoes. The Indians termed it Wapatoo. The starchy tubers were a main vegetable for the Indians who gathered them by wading in the water and loosening them with their feet. Raw, they were bitter, but boiled, were sweet and agreeable.

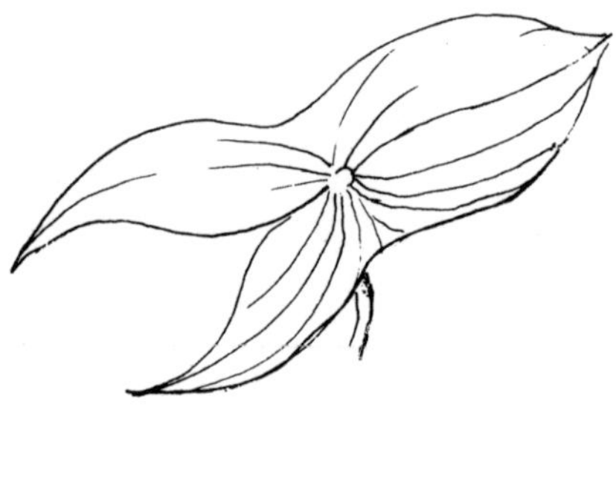

BLACK NIGHTSHADE

Solanum nigrum Solanaceae

The purple flowers are followed by black berries which are considered poisonous. Culpeper wrote: "The whole plant is of a waterish, insipid taste, but the juice of the berries is somewhat viscous, and of a cooling and binding quality . . . It is used to cool hot inflammations either inwardly or outwardly, being in no way dangerous as most of the nightshades are; yet it must be used moderately. The distilled water of the whole herb is safest to be taken inwardly . . . be sure you do not mistake the Deadly Nightshade for this." *(Atropa belladonna)*

BLACK SNAKEROOT

Cimicifuga racemosa Ranunculaceae

The tall, white, fleecy blossoms of this flower give out a most offensive odor. The root was the part of the plant valued in medicine. The Indians believed in Snakeroot's efficacy in the curing of bites by poisonous snakes.

BLOODROOT

Sanguinaria canadensis Papaveraceae

Known as Red Puccoon to the Indians, from its red juice, which comes from any break in the plant or root. Known as White Puccoon for its star-shaped flower which is often in bloom while April snows surround it. It was a most valuable dye plant to the Indians. In use as a dye a mordant is needed to set the rich colors of bronze, red, or yellow. Garments were dyed, also handiwork including baskets. As a war paint, the Indians' bodies, faces, and tomahawks were painted with the red juice. Porcupine quills were dyed red, and wooden implements yellow. Under the Doctrine of Signature, the red juice denoted its use in cases of anemia. Large doses are emetic and narcotic, but in small doses the juice is a tonic and stimulant. *Sanguinaria canadensis* is listed on the present-day Check List of Native and Introduced Drug Plants in the United States.

Bloodroot

BONESET

Eupatorium perfoliatum **Compositae**

Common names for this plant are Ague-weed, Indian Sage and Thoroughwort. Boneset was a truly New England home remedy whose uses were taught the early settlers by the Indians. The growth of its leaves led to the belief of its sagacity to mend broken bones or to aid in "break bone fever." The leaves have the appearance of growing right through the stalk. This comes under the Doctrine of Signature. The mark of this belief is either in the Plant's name, color, or in some peculiarity in its growth, or its shape, which led to the belief of a specific cure.

Boneset

BOUNCING BET

Saponaria officinalis Caryophyllaceae

This herb was introduced at an early date because of its medicinal value. A decoction of the leaves was a remedy for jaundice, gout, or rheumatism; also valuable in the treatment of skin irritations. Known as Soapwort, Scourwort and Latherwort. The Latin *Sapo* means soap. The juice of the leaves froths up into a fine soapy lather which was useful both to the Indians and the first settlers.

BULRUSH

Scirpus spp. Cyperaceae

For use in weaving, bulrushes were collected by the Indian women in canoes after the rushes had reached their maturity either in late June or early July. The rushes were bleached and dried before being woven. Baskets of various shapes and sizes, holding as much as four bushels or as little as a pint, were woven of bulrushes. Mats also were completed to sit on, to line their dwellings, or to hold the shelled corn, berries, etc., which were to be dried and then stored.

BURDOCK

Arctium minus Compositae

Its local names of Beggars' Buttons, Stick Button, Clot Bur and Horse-Burr take note of the persistent clinging burs of this introduced herb. Medicinally, it purified the blood, both the root and seed being made into a decoction. The leaves, in the form of a poultice, brought relief to bruises, tumors and other swellings. It was also considered to be an antidote against the bites of snakes and mad dogs. The root appears on the present-day Check List of Native and Introduced Drug Plants in the United States.

BUTTERFLY WEED

Ascelepias tuberosa Asclepiadaceae

Also known as Pleurisy Root, this plant is one of the most decorative of our native plants with its masses of bright orange flowers. To the Indians the root was a cure for pleurisy. It comes under the definition of a Simple, an herb known for one outstanding medical virtue. A crude sugar was extracted from the flowers by the natives.

Butterfly Weed

CATNIP (or Catmint)

Nepeta cataria Labiatae

Another aromatic herb, beloved by cats, first grew wild, in medieval times, in the Tuscan town of Nept. The leaves made a favorite tea drunk before the teas of the Far East were in common use. It was purported to give courage to the timid and make the weak strong. The tea also was an aid in cases of colds, headache, stomach ache, and pneumonia.

CAT TAIL

Typha latifolia Typhaceae

Cat Tail was native both to England and New England. The outer leaves and cover of the young cat tail were stripped to reveal the slender stem which has a sweetish taste. The root, which is starchy, was preferred as a food by the Indians and could be eaten raw. Coarse mats were woven by the Indian women and these framed and lined the sides of their dwellings. The barrels of the early settlers were made watertight by the dried leaves; mattresses were stuffed with them.

CHICORY

Cichorium intybus Compositae

In England called Succory, this herb is closely related to the dandelion whose leaves, in early spring, it resembles. The tender young leaves were valued as a spring salad. Medicinally, the distilled water of the flowers aided inflammation of the eyes and dimness of sight.

CLEAVERS

Galium aparine Rubiaceae

Cleavers are native to Europe, Britain and the eastern half of the United States and Canada. They were also known as Goose-Grass or Clivers. The plant had many common names, some of which are Catchweed, Beggar-lice, Burhead, Cling-rascal and Stick-a-back. All these names tell us of the formidable burs which cover the plant in the fall. They stick to the fur of animals, the clothing of man, and are all but impossible to remove. Not of economic value to the Indians, the English would have known of its virtue as a medicine to aid jaundice, "taken in broth, to keep them lean and lank that are apt to grow fat," and the juice of the leaves, bruised, stopped the bleeding wounds.

CLOVERS

Red – *Trifolium pratense* Leguminosae
White or Dutch – *T. repens*
White or Sweet – *Melilotus albus*

Bees are attracted to clover, whose flower heads made a delicious mild honey. Medicinally, clover has long been valued as an antispasmodic expectorant and, in the form of an ointment, was applied to ulcers.

COLUMBINE (Wild)

Aquilegia canadensis Ranunculaceae

John Josselyn notes of columbine, which is native both to New England and England, "Columbines, of a flesh color, growing upon rocks." An infusion of the seeds was helpful for jaundice, and a relief for sore mouths or throats.

COW PARSNIP

Heracleum lanatum Umbelliferae

The tip of this plant was a remedy for faintings and convulsions utilized by the Indians; a substitute for salt; and a food.

DAISY (Wild)

Chrysanthemum leucanthemum Compositae

The Indians employed the daisy as a cure for fevers. The English found "The greater wild daisy is a wound herb of good respect, often used in these drinks and salves that are for wounds, either inward or outward."

DANDELION

Taraxacum officinale Compositae

Dandelion was a "Plant sprung up in New England." It was also common in England. The Indians made a tea from the leaves and ate the roots. As a medicine, it was a well known home remedy for cleansing the system. The root, fresh or dried, being the part employed. Culpeper speaks of its "opening and cleansing quality." As a culinary herb, it was cooked in broth, and used in salads.

DOCK (Broad Leaved)

Rumex obtusifolius Polygonaceae

Dock leaves have a soothing principle which made them good as a dressing for blisters, and cases of burns and scalds. It was, and is, very useful in taking out the torment of the nettle's sting.

EVENING PRIMROSE

Oenothera biennis Onagraceae

Listed by John Josselyn as a "Plant proper to the Country:" the bark and leaves were valued for their astringent and sedative qualities.

FERNS

Cinnamon – *Osmunda cinnamonea* Osmundaceae
Hayscented – *Dennstaedtia punctilobula* Polypodiaceae
Interrupted – *Osmunda claytoniana* Osmundaceae
Maiden Hair – *Adiantum pedatum* Polypodiaceae
New York – *Dryopteris noveboracensis* Polypodiaceae
Polypody – *Polypodium virginianum* Polypodiaceae
Royal – *Osmunda regalis* Osmundaceae

"Fern" is mentioned in a general way by John Josselyn. The Indians enjoyed boiling the tender young fronds of Maiden Hair fern, of which fern John Josselyn wrote, "Maiden Hair or *Cappellus veneris verus* which ordinarily is half a yard in height. The apothecaries for shame now will substitute Wall Rue no more for Maiden Hair, since it grows in abundance in New England, from whence they may have a good store." Medicinally, the Indians and early settlers found a decoction of this fern useful in the treatment of chills, fevers and difficult breathing. Polypody was found by the English to be useful as a purge and royal fern was a good wound herb.

FLAGGES

Blue Flag – *Iris versicolor* Iridaceae
Slender Blue Flag – *I. prismatica* Iridaceae
Sweet Flag – *Acorus calamus* Araceae

William Wood wrote: "In summer they (Indians) gather flagges, of which they make mats for houses." Portable mats of flagges were widely used as lodge coverings. Of blue flag John Josselyn noted: "Blew Flower-de-luce; the roots are not knobby, but long and straight, and very white with a multitude of strings. It is excellent for to provoke vomiting, and for Bruises on the Feet or Face. They flower in June, and grow upon dry sandy hills as well as low, wet grounds." Synonyms of this flag are Poison Flag, Flag Lily, Liver Lily, Snake Lily, Marsh Iris or American Fleur-de-Lis. The root must be treated with great caution. The Indians employed it as a cathartic and shared their knowledge with the early colonists. The root appears on the Check List of Native and Introduced Drug Plants in the United States. A mixture of the root of sweet flag, plus powdered willow bark was smoked by the Indians; medicinally it was a remedy for fever, cough or toothache. A paste of the root, rubbed on the face was purported to make an Indian warrior brave. Native to England and New England, sweet flag was also known as Sweet Grass, Sweet Rush and Sweet Sedge. The earliest settlers strewed the sweetly aromatic flag on their floors.

Sweet Flag

FLAX (Wild)

Linum virginianum Linaceae

There are several references to the finding of wild flax by the earliest voyagers and early settlers: "18th December. Many kinds of herbs we found here in winter, and excellent strong kind of flax and hemp." "We had also of them flax, wherewith they make many strings and cords but it is not so bright of color as ours in England. I am persuaded they have great store growing up in the main." Cultivated flax (*Linum usitatissimum*) was well known to the English. The tough fibers of the stalks were the source of linen thread. Medicinally, Culpeper wrote: "The seed, which is usually called Linseed is of great use against inflammations, tumours, and in posthumes, and is frequently put into fomentations and cataplasms, for these purposes."

FOOL'S PARSLEY

Aethusa cynapium Umbelliferae

"There is also great store of Parsley, and divers other wholesome herbs, both for profit and pleasure." This is one of the mystery plants which the early voyagers, along our coasts, misnamed. Fool's Parsley might well be taken for the true as the leaves are similar, but fool's parsley leaves, when bruised, discharge a most disagreeable odor and its synonym of dog poison indicates its poisonous qualities. Animals have died from eating it and most creatures know instinctively that it must not be touched.

GENTIAN

Gentiana spp. Gentianaceae

John Josselyn called it "Autumn Bell Flower" when he found it in New England. There were many varieties of gentian in England and America. Its medicinal uses by the English in the seventeenth century were: "These resist putrefacations and poison, a more sure remedy cannot be found to prevent the pestilence than it is; it strengthens the stomach exceedingly, helps digestion, comforts the heart, and preserves it against faintings and swoonings." – *Culpeper's Complete Herbal.*

GERANIUM

Herb Robert – *Geranium robertianum* Geraniaceae
Wild Geranium – *G. maculatum*

The wild geranium is an attractive low growing plant with pale pink flowers and deeply cut leaves. It prefers low ground and a shady spot. Its many common names include Dove's Foot, Crow Foot, Wild Alum Root, Spotted Geranium and Crane's Bill. Both the natives and the early settlers knew of the root's efficacy as an astringent and as a cure for dysentary or as a general tonic. Herb Robert has purple rose flowers growing out from the slender weak stem which reach up to eighteen inches. The plant has a disagreeable odor at any time but more so when its leaves or stems are crushed. The stems are green only when young and it receives its common name of Red Shanks as they mature. As the leaves show crimson stains, it earns the name of Dragon's Blood. Another common name is Red Robin. Native to England and New England, it was once known as a cure for the plague. Culpeper wrote that "All geraniums are vulnaries but this herb more particularly so."

GLASSWORT

Salicornia europaea **Chenopodiaceae**

John Josselyn wrote of this plant: "Glass-wort, here called Berrilia, it grows abundantly in salt marshes." "Glass-wort, a little quantity of this plant you may take for the Dropsie but be very careful that you take not too much, for it worketh impetiously." Glasswort contains soda. It was found that, mixed with sand, one could produce both glass and soap.

GOLDENROD

Solidago spp. **Compositae**

Native to America, there are over a hundred different species. It was also known of in England where it had been introduced by the Crusaders. The name *solidago* means to make whole and was given to goldenrod because of its healing properties. It had a great repute as a wound herb used externally or internally. Gerard wrote that "It is extolled above all other herbs for stopping of blood in bleeding wounds." The natives brewed a tea from the leaves and a yellow dye was obtained from the flowers and stem.

GOLD THREAD

Coptis groenlandica Ranunculaceae

The twisted mass of yellow roots gave this tiny plant its name. Another name, Canker-root, indicated one medicinal use; the bitter-tasting yellow roots were brewed into a tea to cure canker of the mouth, and it was also a powerful spring tonic. The Indians valued the herb for the same remedies and also employed it to ease a sore throat and aid their babies during the time of teething. They also obtained from the root a yellow dye.

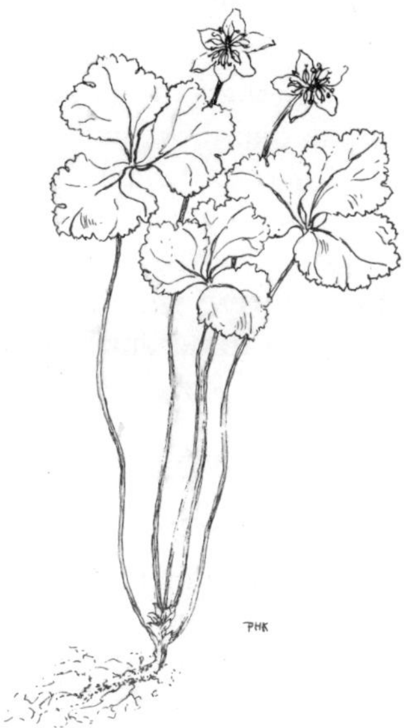

Gold Thread

GOURD

Lagenaria sicerarea Cucurbitaceae

The *lagenaria* originated in Southwest Asia and from there was carried across the Pacific to South America. From there it spread to North America, appearing at some time before maize. Not a food, it was an indispensable part of Indian economy. From the gourds were fashioned dishes, dippers, jugs, bowls, rattles, containers for seed, ornaments to wear, ceremonial masks, water bottles, and floats for nets and rafts.

GRASSES

Marsh Grass – *Spartina pectinata* Gramineae
Silk Grass – *Agrostis hyemalis*
Sweet Grass – *Hierochloe odorata*

The Indians wove baskets of many of the native grasses. Sweet Grass was the one most widely used, probably because of its lovely fragrance when dry. Baskets were woven of this grass; trimmings and ornaments of the grass, coiled, were decorative motifs for many articles. Bowls, mats, dishes and trays were among the items made, and it was powdered as a perfume and incense for ceremonial rites.

GROUND CHERRY

Physalis pubescens or *virginiana* Solanaceae

Ground Cherry is a member of the nightshade family and a relative of the tomato. Its sweet, edible, yellow fruits are contained in straw-colored husks. The Indians knew this plant, and the early settlers enjoyed the fruit raw, or cooked in sauces and pastries.

GROUND IVY

Glechoma hederacea Labiatae

This low-growing, aromatic herb, also known as Gill-over-the-Ground, is closely allied to catnip. It had many medicinal values. John Gerard wrote that "Boiled in mutton broth it helpeth weak and akeing backs." It was also an aid to coughs, headache and inflammation of the eyes. It was called Ale Hoof and Tunhoof as it was incorporated in the brewing of beer to clarify malt liquors.

GROUND NUT

Apios tuberosa Leguminosae

"Also, in every island, and almost every part of every island, are great store of ground nuts, forty together on a string, some of them as big as hen's eggs: they grow not two inches under ground: the which nuts we found to be as good as potatoes." – John Brereton, 1602. John Josselyn wrote of ground nut: "For to eat out the proud flesh they (Indians) take a kind of earth nut boyled and stamped." Ground Nuts were one of the most important native foods to the Indians. They shared their knowledge with the Pilgrims which well may have helped to save lives during that first harsh winter. William Bradford wrote: "At last most of them left their dwellings and scattered up and down in the woods and by the water sides, where they could find ground nuts and clams, here six there ten," "and this was their living in the summer time, till God sent them better; and in winter they were helped with ground nuts and fowl."

HEAL ALL

Prunella vulgaris Labiatae

Heal All, All Heal, or Self Heal's astringent properties make it an excellent gargle. Its other common names of Carpenter Herb, Stickle Wort and Hook Heal indicate its true value as a healing herb. Applied externally in the form of an unguent or plaster it quickly sealed up cuts, and the juice, drunk in the form of a syrup, healed wounds and ulcers.

HELLEBORE (Green)

Veratrum viride Liliaceae

John Josselyn wrote: "White Hellebore, which is the first plant that springs up in the Country, and the first that withers; it grows in deep black mold and wet, in such abundance that you may in a small compass gather whole cartloads of it." It grows near skunk cabbage and is often mistaken for it. John Josselyn continues: "The Indians cure their Wounds with it, annointing the Wound first with Raccoons greese, or Wild-Cats greese, and strewing upon it the powder of the roots; and for Aches they scarifie the grieved part, and annoint it with one of the foresaid Oyles, then strew upon it the powder: The powder of the root put into a hollow tooth is good for Tooth-ach: The Root sliced thin and boyled in Vinegar, is very good against Herpes Milliaris." It was called poke by the Indians. In spite of the fact that many sources say that the herb must be used with extreme caution, the natives and early settlers found it a delicious pot herb.

INDIAN CUCUMBER ROOT

Medeola virginiana Liliaceae

The Indians ate the root which is tuberous and somewhat the shape of a cucumber.

INDIAN HEMP

Apocynum cannabinum Apocynaceae

"There is Hempe that naturally groweth finer than our Hempe of England," 1622. "In summer they (Indians) gather flagges . . . and Hempe and Rushes, with dyeing stuff of which they make curious baskets." 1629. Hemp was utilized by the natives for thread, bags and sacks which held up to five or six bushels, and bone fish hooks were kept in place by hemp.

INDIAN PIPE

Monotropa uniflora Pyrolaceae

Known as Corpse Plant for its white eerie, wax-like appearance, Indian pipe grows in leaf mold in deep woods. The Indians employed the plant medicinally in a lotion to heal the eyes.

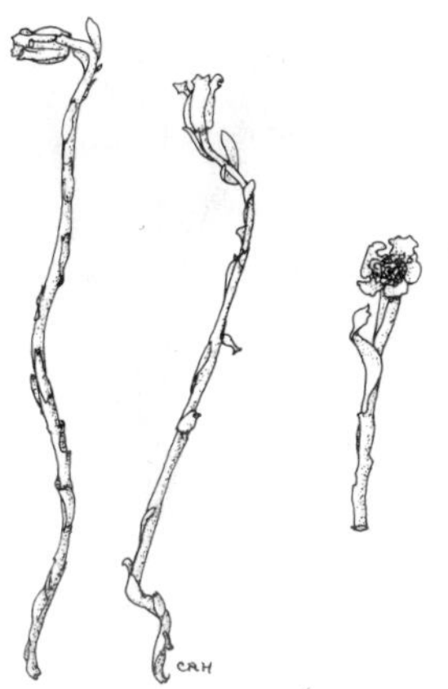

Indian Pipes

JACK-IN-THE-PULPIT

Arisaema triphyllum **Araceae**

Although the roots, eaten fresh, are poisonous, the Indians cooked and ate the starchy, rather tasteless roots. An application of the dried and grated roots was a remedy for headache.

Jack-in-the-Pulpit

JEWEL WEED

Impatiens pallida or *biflora* Geraniaceae

Common names of this plant are Pale Touch-me-not, Balsam Weed, or Speckled Jewels. The Indians taught the early colonists its use as a cure for poison ivy. John Josselyn wrote of it: "This Plant the Humming Bird feedeth upon, it groweth likewise in wet grounds, and is not at its full growth till July, and then it is two cubits high and better, The Leaves, thin and of a pale green colour, some of them as big as a nettle leaf, it spreads into many branches, knotting at the setting one, and of a bright yellow colour, speckled with a deeper yellow as it were shadowed, the Stalkes are as hollow as a Kix, and so are the roots, which are transparent, very tender, and full of yellowish juice. The Indians made use of it for aches, being bruised between two stones and laid to cold, but made (after the English manner) into an unguent with Hog's Grease, there is not a more sovereign remedy for bruises of what kind so ever; and for Aches upon Stroaks."

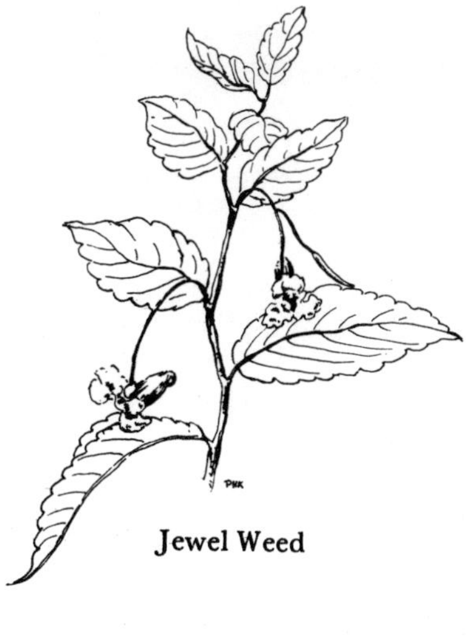

Jewel Weed

JOE-PYE WEED

Eupatorium purpureum Compositae

Common names for joe-pye weed are Trumpet Weed, Purple Thoroughwort, Gravel or Kidney-root and Fall or Purple Boneset. Joe-Pye was an Indian medicine man who showed the first colonists the medicinal use of the herb's root as a cure for typhus fever.

LADY'S SLIPPERS

Pink – *Cypripedium acaule* Orchidaceae
Yellow – *C. pubescens*

Called by the Indians Moccasin Flower. To them the root was a remedy as a tonic and a gently nervous stimulant. Its root is included in the Check List of Native and Introduced Drug Plants of the United States.

LIFE ROOT

Senecio aureus Compositae

The Indians applied the root and whole herb medicinally. Known as Golden Ragwort, or Squaw Weed, it formerly enjoyed a reputation as a female regulator.

LILIES

Yellow Water or Pond Lily – *Nuphar advena* Nymphaeaceae
Sweet Scented Water Lily – *Nymphaea odorata*

Water Lilies were an important food to the Indians. John Josselyn wrote: "Water Lily, with yellow flowers, the Indians eat the roots, which are long a boiling, they taste like the Liver of a Sheep, the Moose Deer feed much upon them, at which time the Indians kill them, when their heads are under water." Not only the roots were eaten, the seeds were boiled, roasted or eaten raw and even the blossoms were cooked. The early settlers also enjoyed them. As a medicine: "An Indian, whose knee was bruised with a fall, and the Skin and Flesh stripped down to the middle of the Calf of his leg; cured himself with Water Lily roots boyled and stamped." "They (the Indians) apply to the Sore the roots of Water Lilies boiled and stamped betwixt two stones to a Playster." – John Josselyn.

LIVERWORT

Hepatica americana Ranunculaceae

"Noble Liverwort, one sort with white flowers, the other with blue." – John Josselyn. The leaves, pounded to a pulp, and made into a decoction, was thought by the Indians to rid themselves of the nightmare. To the first settlers the liver-shaped leaves indicated that the herb would cure any ailment of the liver.

LOOSESTRIFE (Purple)

Lythrum salicaria Lythraceae

This spectacular herb, introduced at an early date, is described by John Josselyn: "Lysimachus or Loose Strife, it grows in dry grounds in the open Sun four foot high, flowers from the middle of the plant to the top. The flowers purple, standing upon a small sheath or cod, which when it is ripe, breaks and puts forth a white silken doun, the stalk is red, and as big as one's Finger." The distilled water, in the form of a lotion, preserved sight and cured sore eyes. An ointment of the herb was an aid to wounds.

MARSH MALLOW

Althaea officinalis Malvaceae

This naturalized herb has tiny pink flowers and is found only in the salt marshes along the coast from Massachusetts to New York. Of value medicinally because of its mucilaginous root, it was a cure for coughs and sore throats. The roots were edible.

MAY APPLE

Podophyllum peltatum Berberidaceae

Also called Hog Apple, American Mandrake or Wild Lemon, the may apple was best known for its medicinal use, although the lemon-shaped fruits called "apples" were eaten and had a sweet but rather insipid taste. The Indians showed the early settlers its great value as a medicinal herb. The root was the part employed and in overdoses could be poisonous. Used with care, it was invaluable in promoting discharge of bile, had great power in action upon liver and bowels, stimulated glands to healthy action and exercised an excellent influence on every part of the system. The root is on the Check List of Native and Introduced Drug Plants of the United States.

MAY WEED

Anthemis cotula Compositae

Of this introduced plant, John Josselyn wrote: "May Weed, excellent for the Mother: some of our English housewives call it Iron Wort and make a good unguent for old sores." A weed of the fields, it was hated by farmers as the leaves and seeds blistered the hands, ankles and feet as they worked on their crops.

MILKWEED

Asclepias syriaca Asclepiadaceae

One of the most beautiful sights in nature is the opening of the milkweed pods and the subsequent cascading of the white silky hairs with their attendant seeds which sail in the wind to great distances. The young stalks and blossoms were boiled and eaten by both the Indians and the early settlers. A letter written by John Winthrop speaks of pods of milkweed "used to stuff pillows and cushions."

MINTS

Applemint – *Mentha rotundifolia* Labiatae
Peppermint – *M. piperita*
Spearmint – *M. spicata*
Wild Mint – *M. canadensis*

Apple, Pepper and Spearmint were introduced at an early date. It is known that Spearmint was planted in Elder William Brewster's garden before 1630. The English, before they came to New England were accustomed to strew mints in churches, and adding them to baths. The smell of mint stirred up the mind; and was recommended to students "oft to smell." They whitened teeth. The mints flavored sauces, vinegar and jelly; added flavor to meat dishes. Culpeper wrote of them: "All mints are astringent and of warm subtle parts; great strengtheners of the stomach. Their fragrance betokens them caphalica; they effectually take off nauseousness and retchings . . . the simple water given to children, removes the gripes but these virtues more particularly belong to Spear and Peppermint." Our native wild mint grows from New Brunswick to Virginia and westward. It is similar in odor and appearance to pennyroyal (*Mentha pulegium*).

MOSSES

Irish Moss – *Chondrus crispus* Algae
Sphagnum Moss – *Sphagnum spp.* Sphagnaceae

Irish Moss grows along the sea coast, and was invaluable medicinally. It contains iodine and has mucilagenous qualities which made it a panacea for sore throats. The Indians found that Sphagnum Moss, when soaked in water, grew to twice its normal size and was antiseptic and cooling. John Josselyn wrote the following: "The Indians make use of the Moss boiled in Spring water, for Stabbs, pouring in the Liquor, and applying the boiled Moss well stamp'd or beaten betwixt two stones. Of the moss that grows at the roots of the White Oak the Indians made a strong decoction, with which they help their Papouses or young children's scall'd Heads."

MULLEIN

Verbascum thapsus Scrophulariaceae

The spectacular six-foot yellow flower and velvety leaves of this introduced biennial gave it many common names. High-taper, because a tallowed torch was made of the flower spike, and Velvet Plant, Flannel Leaf and Aarons-rod. Also Bullock's Lungwort as it cured diseases of the lung in man or beast. The flowers were a source of a bright yellow dye. Both the Indians and early settlers valued mullein's leaves and flowers for the relief of chest complaints. The herb helped coughs and a tea, brewed from the leaves, was given in cases of fever.

MUSTARD (Black)

Brassica nigra Cruciferae

Black Mustard seeds, pulverized in a mortar, have been used since Roman days as a condiment. In England John Parkinson wrote: "The seeds ground between two stones, fitted for the purpose, and called a Queene, with some good vinegar added unto it, to make it liquid and running, is that kinde of mustard that is usually made of all sorts to serve as sauce both for fish and flesh." In 1615 we are advised: "For syatica, use a plaster of mustard." This herb was introduced at an early date into New England.

NETTLE

Urtica dioica Urticaceae

John Josselyn wrote: "Nettlesstinging, which was the first plant taken notice of." The long, tough, white shredded fibers procured from the dried stalks had many uses for the Indians. Fish nets were made, snares for rabbits, traps for Otter, thread, and bow strings. Also common to England, it was thought to have been introduced by the Romans. The English brewed a tea of the herb, which purified the blood. Also "The Buds are very tender crimae, a little bruised, are by some eaten raw, or by others boil'd, especially in Spring potage, with other Herbs." – John Evelyn.

ORACH (Green)

Atriplex spp. Chenopodiaceae

John Josselyn called this a "Plant Sprung up in New England, Wild Arrach." It was indigenous to England and New England. The English benefited from this annual in the following manner: "Orach, Atriplex is cooling, allays the Betuit Humor: Being set over the Fire neither this, nor lettuce, needs any other Water than their own moisture to boil them in without expression: The tender leaves are Mingl'd with other cold salletting: but 'tis better in Pottage." – John Evelyn.

PENNYROYAL (American)

Hedeoma pulegioides Labiatae

"Also divers excellent pot herbs grow abundantly among the grasse . . . and pennyroyal" – *New England's Plantation*. Another account mentions pennyroyal and calls it a "sallet herb." Known as Mock Pennyroyal, Tickweed and Squaw Mint, American pennyroyal is a highly aromatic herb which is very similar to the European pennyroyal *(Mentha pulegium)*. The Indians and first colonists brewed a tea of this herb, as a cure for measles or whooping cough.

PEPPER ROOT

Dentaria laciniata Cruciferae

Also called Crinkle Root, Toothwort, Crow's Foot and Wild Horseradish, the Indians ate the pungent root.

PLANTAIN

Plantago major Plantaginaceae

"Plantain, which the Indians call English-man's Foot, as though produced by their treading." The leaf does resemble a man's foot, and *plantago* means sole of the foot. This weed has many common names including Way Bread, Devil's Shoestring, Bird Seed, Snakeweed, and Rib or Ripple Grass. Plantain leaves, applied, were a fine wound herb used by both the English and the Indians. The Indians also employed it as an antidote against the poison of snake bites, and the English found it a good remedy as a relief from stings and bruises.

POISON IVY

Rhus toxicodendron Anacardiaceae

The juice of poison ivy was invaluable to the Indians as a permanent and deep black dye. The same juice has been employed as an ink.

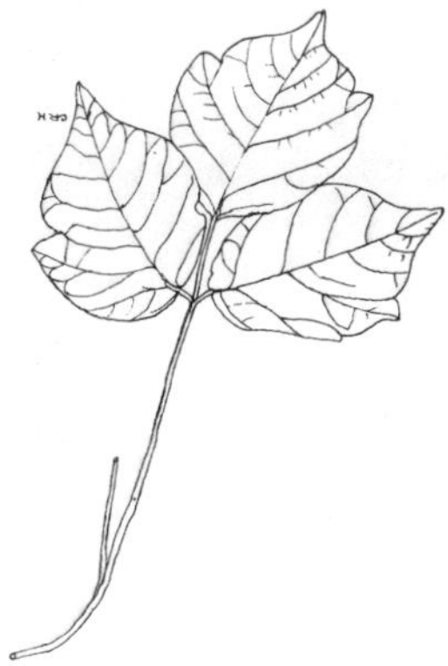

Poison Ivy

PURSLANE

Portulaca oleracea Portulacaceae

"They brought also some Purslane, which grows in great quantities among the Indian corn and of which they make no more account than of the weeds." – *Sailor's Narratives of Voyages Along the New England Coast, 1524-1624*. The Indians made a mush by grinding the seeds to a flour. Well known to the early settlers, as it was also common in England, purslane was highly regarded as a pot herb, and much eaten in salads.

RATTLESNAKE ROOT (Cankerweed or White Lettuce)

Prenanthes alba Compositae

John Brereton, on a voyage to these shores in 1602, wrote the following: "Thirdly, the country being very full of woods and wildernesses doth also much abound with snakes and serpents called rattlesnakes that have rattles in their tail, that will not fly from man as others will, but will flye upon him, and sting him so mortally, that hee will dye within a quarter of an houre after, except the parties stinged have about him some of the root of an herb called snake-weed to bite on, and then hee shall receive no harme."

ROCKFOIL

Saxifrage virginiensis Saxifragaceae

As rockfoil grows upon rocks it was thought that "there is no better medicine than this to break stone." Taken in beer, the herb was a panacea against raging fevers.

SABATIA

Sabatia kennedyana Gentianaceae

Of interest only as it is a plant peculiar to the Plymouth, Massachusetts area. It is believed that the Pilgrims, in the summer of 1622, discovered it on a Sunday and named it sabatia. Also known as Rose of Plymouth.

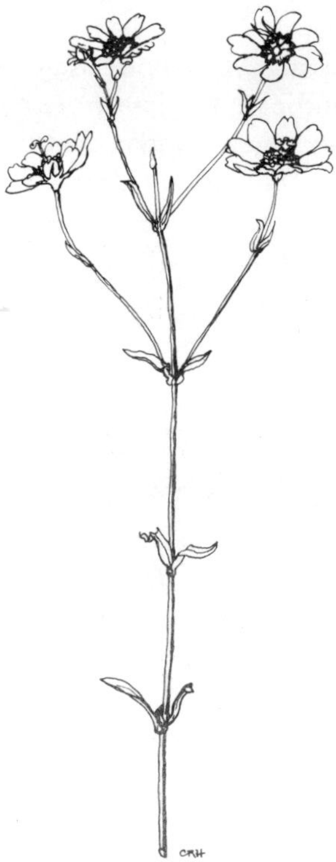

Sabatia

ST. JOHN'S WORT

Hypericum perforatum Guttiferae

This introduced herb comes under the Doctrine of Signature. The leaves, when held up to the light, show tiny perforations. This, plus the fact that the stem held a red juice, led to the belief that St. John's Wort was a superior wound plant.

SCARLET PIMPERNEL

Anagalis arvensis Primulaceae

An annual whose names of Poor Man's Weather Glass or Shepherd's Clock indicate the habit of the flowers to close whenever the sky becomes cloudy. Native also to England, the English knew of its value in drawing out splinters and thorns, and internally or externally as a cure for the bite of all poisonous and rabid creatures.

Scarlet Pimpernel

SHEPHERD'S PURSE

Capsella Bursa-pastoris Cruciferae

This herb was listed by John Josselyn as follows: "Of such Plants as have sprung up since the English planted and kept cattle in New England." Shepherd's Purse has many common names; Case Weed, Pick Purse, Mother's Heart. The last named refers to the shape of the seed pods and the heart-shaped seeds contained in them. The styptic quality of the herb "stoppeth blood," and an ointment made of the herb was excellent for wounds, especially those in the head. Shepherd's Purse appears on the Check List of Native and Introduced Drug Plants in the United States.

SHIN LEAF

Pyrola elliptica Pyrolaceae

Shin Leaf received its common name from the first colonists who applied the leaves as a "shin plaster" for bruises.

SKUNK CABBAGE

Symplocarpus foetidus Araceae

John Josselyn wrote: "The whole plant sents as strong as a Fox, it continues till August." This foetid plant, growing in swamp areas, starts to unfold its beautiful leaves long before any other plant in the spring. The Indians called it "Skota" and found that the ground and powdered roots made a very fine styptic. As a culinary herb, they baked the root, extracted the juice, and the root provided a bread stuff.

SNAKEHEAD

Chelone glabra Scrophulariaceae

Also known as Turtle Head as the pink flower closely resembled a snake or turtle head. This herb was a remedy for the Indians in liver complaints. John Josselyn lists it as "Such Plants as are Proper to the Country."

SOLOMON'S SEAL

Polygonatum biflorum Liliaceae

Growing also in England, solomon's seal received its name from the fact that its root resembles a seal. Medicinally, it comes under the Doctrine of Signature, as it was purported to be excellent for "sealing" up wounds and broken bones.

SORREL (White or True Wood)

Oxalis montana Oxalidaceae

References to finding sorrel in New England lists it as a pot and salad herb. The juice is very acid; oxalis being the Greek word for sour. The Indians cooked the leaves, and, with sweetening added, ate it as a dessert.

SPEEDWELL (or Brooklime)

Veronica americana Scrophulariaceae

There is some difference of opinion as to whether this plant is naturalized or native. The American speedwell grows over most of the United States along banks or streams, ponds and rivers, and is very similar in appearance to forget-me-not. Speedwell was known as a medicinal herb; a tea served as a tonic; and it was given in cases of fever.

Speedwell

SUNFLOWER

Helianthus annuus **Compositae**

John Josselyn wrote of sunflower: "This Plant is taken by our Simplists to be a kind of Golden Rod, by others for Sarazens Confound. I judge it to be a kind of small Sun Flower, or Marygold of the West Indies; the Root is brown and slender, a foot and a half in Length, running a slope under the upper face of the Earth, with some strings here and there, the stalk as big as the steal of a Tobacco pipe, full of pith, commonly brownish, sometimes purple, three or four feet high, the Leaves grow at a distance one against another, rough hard, green above, and gray underneath, slightly snipt and the ribs appear most on the back side of the Leaf, the Flower is of a bright yellow, with little yellow cups in the midst, as in the Marygold of Peru, with black threads on them with yellow pointels, the Flower spreads itself abroad out of a cup made up of many green beards, not unlike a Thistle: Within a handful of the top of the stalk, when the Flower is fallen, growes an excrense or knob as big as a Walnut, which being broken yieldeth a kind of Turpentine or rather Rosein." There are at least six different *Helianthus* but *annuus* was the one brought under cultivation, along the Atlantic coast by the Indians. The Sunflower seeds were an important food to the Indians. They cooked both a bread and a broth of them. The boiled seeds made a hair oil, chewing the seeds was an antidote for snake bite, and the flower heads were employed in cases of lung trouble. The stalks furnished a fiber, the leaves fodder for animals and the flowers produced a yellow dye.

TANSY

Tanacetum vulgare **Compositae**

Tansy was brought to New England by the early colonists where it "flourisheth notably." Very soon it made its escape from the garden plots and into the nearby fields. The common name of Bitter Buttons indicates its bitterness, which we find makes it not very palatable but it was, formerly, a culinary herb. A "Tansy Pudding" was concocted from the young leaves mixed with eggs and eaten during Lent as a welcome change from a salt fish diet. It was incorporated, sparingly, into salads. As a strewing herb, its strong bitter odor kept flies away, and meat rubbed with the herb helped prevent decay. As a cosmetic it aided the removal of freckles. Medicinally, it was a cure for the plague; Tansy tea was helpful for colds and rheumatism.

THYME (Wild)

Thymus serpyllum Labiatae

Wild Thyme escaped from the first gardens in New England. Cultivated in England before the middle of the sixteenth century, the English made use of the herb in a variety of ways. The clean and refreshing scent made it a lovely strewing herb. It was a part of salads, and an addition to broth; and made a "sawce for divers sorts both fish and flesh." Medicinally Culpeper wrote: "It is a strengthener of the lungs; a good remedy for the chin cough in children." The aromatic tea refreshed people in weak health. The blossoms attract bees by the thousands.

TOAD FLAX

Linaria vulgaris Scrophulariaceae

This introduced herb is also known as Butter and Eggs because of the shape and color of the bright flower spikes which appear at the end of the stem. Valued by the English as a medicine, John Gerard wrote that: "the decoctions openeth the stopping of the liver and spleen, and is singularly good against the jaundice which is of long continuance."

TOBACCO

Nicotiana tabacum Solanaceae

There are many references to the planting of tobacco by the Indians and of their sharing it with the early settlers: "Before reaching their cabins, we entered a field planted with Indian corn . . . some tobacco." "Captain Standish and Isaac Allerton went venturously, who were welcomed of him after their manner: he gave them three or four groundnuts and some tobacco." "He (The Great Sagamore) told me also that another time there was a man who had a large quantity of tobacco (a plant from which they obtain what they smoke)." (Nauset) – "They gave us a large quantity of tobacco which they dry and then reduce to Powder." The Indians found that tobacco strengthened them and took away weariness. It was considered to be a sacred plant and was a part of every religious ceremony whether it be a puberty rite, a funeral, war expedition, or harvest festival. Tobacco was thought to ward off disease and quiet fear. It established friendship. It was the only plant cultivated by the Indians which the men tended, the women usually took care of the gardens.

TRAILING ARBUTUS

Epigaea repens Ericaceae

One of the earliest blooming and best loved of our native flowers with its deep green leaves and so fragrant white or pink flower, which grows deep in the woods. It was called Mayflower by the Pilgrims who saw in the rise of the new leaves over the brown of last year's foliage a parallel to their own rise over great hardships.

Trailing Arbutus

TRILLIUM

Trillium erectum Liliaceae

Also known as Wake Robin, Indian Balm, Ground Lily, Birth Root, Cough Root or Snakebit. Both Indians and early settlers employed the herb medicinally. It is astringent and antiseptic. A poultice, applied externally, helped skin disease and retarded gangrene.

VIOLET

Viola spp. **Violaceae**

The English must have been delighted to find so many different kinds of violets here as they were one of their best loved flowers. "Violets are the spring's chiefe flower for beauty, smell and use." Violets were an ingredient of cosmetics, sweet waters and a strewing herb. They were incorporated in salads, soups, sweet meats, and a wine was made of violets. Medicinally, syrup, conserve, and a paste were highly recommended for "delicate people." The flowers were an antidote for pain; the herb, bound to the forehead, induced sleep and the flowers, fresh or dried "are effectual in the pleurisy, and all diseases of the lungs."

WATER PARSNIP

Sium suave — Umbelliferae

"The Eight and twentieth we entered counsel about abode and plantation which was concluded to be in the West part of Elizabeth's Island . . . the herbs and roots are . . . skirring without count." Water Parsnip is a native plant found in swampy places throughout Canada and the United States. It is a close relative to the skirret of England *(Sium sisarum)* and might well have been mistaken for it. Sometimes called indian rhubarb, the root provided a pot herb to the Indians. Listed as poisonous, it was not the only plant so called to be eaten by the Indians.

Water Parsnip

WILD BERGAMOT

Monarda fistulosa Labiatae

This aromatic herb, of the mint family, has flowers which vary from whitish, lavender, purple, yellowish pink, pink rose to a bright scarlet. It was also known as Bee Balm as it is a favorite bee plant; also Oswego Tea, as a tea was brewed of it by the Indians.

WILD CARROT (or Queen Anne's Lace)

Daucus carota Umbelliferae

This biennial, with its fern-like foliage and lovely lacy blossom, is the plant from which the vegetable was derived. A juice of the herb was a remedy for the Indians to relieve aching feet. Known to the English, its medicinal virtues were considered to be the same as carrot, but some believed the wild carrot to be more efficacious. Culpeper recommends: "The seed also of the same worketh the like effect, and is good for dropsy, and those whose bellies are swollen with wind; helpeth the colic, the stone in the kidneys . . . The leaves being applied with honey to running sores or ulcers, do cleanse them."

WILD GINGER

Asarum canadense Aristolochiaceae

The root yielded an aromatic stimulant and tonic; the natives found it an effective flavoring for food.

WILD INDIGO

Baptisia tinctoria Leguminosae

This is a bushy plant covered with bright yellow, pea-shaped flowers during the summer, followed by black seed pods. A relative of indigo, it is a poor substitute for that plant as a dye. The Indians employed it medicinally as a purgative; and, applied it externally, for bruises, cuts and scratches.

WILD SENNA

Cassia marylandica Leguminosae

The root of this plant was of medical use to the Indians as a cathartic, a poultice for sores, and, in the form of a decoction, an aid in cases of fever.

YARROW

Achillea millefolium Compositae

Yarrow was both native and introduced. Yarrow's other common names indicate its uses by the English: Sneezeweed, Nosebleed, Knights Milfoil, and Soldiers' Woundwort. It was a wound herb of great repute. The name Old Man's Pepper refers to its pungent odor and taste. Yarrow appears on the present-day Check List of Native and Introduced Plants of the United States. The Indians found an infusion serviceable for earaches and swollen tissues and held the whole herb in their mouth for toothache.

Vegetables

BEANS

Kidney or Common – *Phaseolus vulgaris* **Leguminosae**
Scarlet Runner – *P. multiflora*

The bean was the second most important food crop of the Indians. In *Champlain's Voyages* there are many references to the Indians growing beans, among them: "From this meal (corn) they make bread, using also beans which they first boil, as they do the Indian corn for soup, so that they may be more esily crushed. Then they mix all together sometimes adding blueberries or dry raspberries." The Pilgrims, in their first explorations of Cape Cod, found corn and beans which they appropriated for their own use. The Pilgrims planted the beans, as did the Indians, near the corn, so that the beans might entwine themselves around the corn stalks.

CORN (or Maize)

Zea mays Gramineae

To the Indians corn was a most important and widely planted food crop. It has no known wild relative, has been cultivated for thousands of years, and probably originated in Mexico. "Their food is generally boiled maize, or Indian corn, they boil it in an earthen pot, which they make in a way different from ours. They bray it also in wooden mortars and reduce it to flour, of which they make cakes." – *Champlain's Voyages*. During the Pilgrims' first explorations of Cape Cod, from the Shallop, they found areas which had been planted with corn, and baskets full of corn. On March 22 "Samoset and Squanto stayed all night with us. They said that within eight or nine days they would come and set corn on the other side of the brook." The Pilgrims did plant twenty acres of corn that first spring and, under the Indians' direction, manured each hill with alewives from Town Brook. "It is almost incredible what great gaine some of our English Planters have had of our Indian corn. Credible persons have assured me, and the partied himselfe avouched the truth of it to me, that of the setting of thirteen gallons of corn hee hath encrease of it 52 hogsheads, every hogshead holding seven bushels of London measure, and every bushel was by him sold and trusted to the Indians for so much beaver as by reckoning will appeare; where you may know how God blessed husbandry in this land. There is not such greate and plentifull eares of corn I suppose anywhere else to bee found but in this country: because also of varieties of colours, as red, blew, and yellow etc., and of one corne there springeth four or five hundred. I have sent you many eares of divers colours that you might see the truth of it." – *New England's Plantation, 1629.*

JERUSALEM ARTICHOKE

Helianthus tuberosus **Compositae**

Jerusalem Artichoke is a perennial species of sunflower with edible tubers resembling potatoes. It was cultivated over a large range by the Indians, and the tubers were boiled or roasted, or sometimes eaten raw. The Indians showed the early colonists how to grow and use this plant. Early in the seventeenth century the plant had been taken to England where it received the name of jerusalem artichoke. Jerusalem is a corruption of "girasole" which is sunflower in Italian, and artichoke must have meant that the plant's taste was reminiscent of artichoke.

Jerusalem Artichoke

MELONS

Musk – *Cucumis melo* Cucurbitaceae
Water – *Citrullus vulgaris*

The Indians "have a good store" of water melons and John Josselyn notes: "It is a large Fruit, but nothing near so big as a pompion, colour smoother and of a sad Grass green rounder or more rightly Sap Green; with some yellowness admixt when ripe; the Seeds are black the flesh or pulpe exceeding juicy." Of musk melon he found them an improvement on the English variety.

ONIONS

Wild Leek – *Allium tricoccum* Liliaceae
Wild Onion – *A. canadense*

The natives and first colonists did not need to cultivate onions, as there was a great plenty of them growing naturally. In *New England's Plantation* we read: "this country aboundeth naturally with store of roots of great variety and good to eat . . . also leeks and onions are ordinarie." John Josselyn adds: "Wild Leeks which the Indians use much to eat with their fish."

PUMPKIN

Cucurbita pepo **Cucurbitaceae**
Squash – *C. maxima*

Pumpkins, also known as pompions or pumpions, and squashes should be considered together as the names were often interchangeable. There are many references to the early discoverers finding the Indians cultivation both in their gardens. Squash, pumpkins, and beans were usually planted in the corn field. John Josselyn wrote: "Squash, pumpkins, and beans were but more truly squonter squashes, a kind of mellon, or rather gourd, for they often times degenerate into gourds; some of these are green, some yellow, some longish like a Gourd, others round like an apple, all of them pleasant food boyled and buttered, and seasoned with spice; but the yellow squash called an Apple Squash, because like an apple, and about the bigness of a pome water is the best kind; they are much eaten by the Indians and the English." Both pumpkin and squash were cooked in much the same manner as today. The Indians also roasted the seeds and pieces of pumpkins were hung on strings and dried for winter use.

BERRIES

BLACKBERRY

Rubus spp. **Rosaceae**

Blackberries were dried in the sun by the Indians for winter use. Sometimes to hasten the process, the berry clusters were laid on flat rocks, near, but not too near, their fire. The English ate the fresh berries and as a medicine the berries had the following virtues: "The leaves and brambles, as well green as dry, are excellent good lotions for sores in the mouth" and "the berries or the flowers are a powerful remedy against the poison of the most venemous serpents." – Culpeper.

BLUEBERRY

High Bush – *Vaccinium corymbosum* **Eircaceae**
Low Bush – *V. angustifolia*

All the early voyagers along the New England coast mention finding blueberries. In England they were known as Billberry, Whortleberry or Hurts, depending on the area of the country in which they grew. John Josselyn wrote: "Bill Berries, two kinds, Black and Sky coloured, which is more frequent. They are very good to allay the burning heat of Feavers, and hot Agues, either in Syrup or Conserve. They usually eat of them put into a Bason, with Milk, and sweetened a little more with Sugar and Spice, or for cold Stomachs in Sack. The Indians dry them in the sun, and sell them to the English by the Bushell, who make use of them instead of Currence, putting them into puddens, both boyled and baked, and into Water Gruel." The Indians dried the berries in great quantities. Dried, they were immune to decay, and ground into a powder they were employed as flavoring for stews. The dried berries were also cooked with the Indians' meat, put into soups and sometimes added to their corn bread.

CURRANT

Black – *Ribes nigrum* Saxifragaceae
Red – *R. sativum*

"Currant bushes are of two kinds, red and black. The black currents which are larger then the red, are reasonable, pleasant in eating." – John Josselyn. And William Wood adds: "In other seasons there bee Gooseberries, Billberries, Raspberries, Currants; which being dried in the Sunne are little inferior to those that our Grocers sell in England." Native here, "currence" were listed among the private ship provisions to be brought over by the passengers on the Mayflower. Along with the other native berries, currants were much eaten by the Indians.

GOOSEBERRY

Ribes grossularia Saxifragaceae

In England gooseberries "young and tender leaves are eaten in sallads," and the fruit was cooked as a sauce for meat. Medicinally the berries were considered "cooling and astringent, creating an appetite and quenching thirst." The English were pleased to find these berries in New England and the Indians ate them as well as the other berries.

GRAPES

Fox or Plum – *Vitis labrusca* Vitaceae
Frost or Winter – *V. vulpina*
Riverside or Sweet Scented – *V. riparia*

There was such an ample supply of grapes growing naturally along the New England coast that the Viking explorers named the country Vinland and Gosnold, in 1603, called one of the islands Martha's Vineyard. William Wood in 1634 wrote: "This tree (Hornebound tree) growing with broad spread Armes, the vines wind their curling branches about them; which vines affoard great store of grapes, which are very big both for the Grape and Cluster, sweet and Good. These be of two sorts, red and white, there is likewise a smaller kind of grape, which groweth in the Islands which is sooner ripe and more delectable." Grapes were eaten; a wine was made, the juice of the unripe fruit, known as verjuice, was a remedy to be applied to sprains and bruises; the sap of the vine was a cure for weak eyes.

RASPBERRY

Rubus idaeus and varieties — Rosaceae

Raspberries were dried by the natives in the same manner as blackberries. Dried raspberries or blueberries were sometimes added to the Indian corn bread. The English ate the berries and cooked a raspberry jelly. Medicinally: "The flowers of Raspis are good bruised with hony, and layde to the inflammations and hoate humours gathered together in the eyes." – *Dodoens Niewe Herball.* Raspberry leaf tea helped women in childbirth.

STRAWBERRY

Fragaria virginiana — Rosaceae

"There are likewise Strawberries in abundance, very large ones, some being two inches about; one may gather halfe a bushell in a forenoon." – William Wood. Native to both England and New England, strawberries were loved and of great benefit to the English. A tea was brewed from the leaves, to which sweet Woodruff leaves were sometimes added. The leaves were a welcome addition to cooling drinks, and the juice of the strawberry was a complexion wash. Winslow, in treating an illness of Massasoit, made a broth of a handful of strawberry leaves, and sliced roots of sassafras. The Indian name for the herb was Wuttahimneash. Mixed with corn meal, the Indians made a strawberry bread; the fruit was employed to make a red dye; and a tea was brewed of the leaves, sassafras root, wintergreen leaves, and sweet birch twigs.

BEARBERRY

Arctostaphylos Uva-ursi Ericaceae

Some of the common names of this prostrate plant, with shiny green leaves, small white flowers, followed by bright red berries, are Mealyberry, Bear's Grape, Foxberry, Rockberry, Crowberry, Hog Cranberry, Sagachomi, and Kinnikinnick. Native to America and Europe, it has been used as a medicinal plant since the thirteenth century. Its properties are astringent, diuretic, and a mild disinfectant. The leaves, dried, were smoked by the Indians.

Bearberry

CRANBERRY

Vaccinium macrocarpon Ericaceae

The Indians cooked the tart berries with maple sugar or honey. John Josselyn wrote: "Cranberry, or Bear Berry, because bears use much to feed upon them, is a small trayling plant that grows in Salt Marshes that are over grown with moss; the tender branches (which are reddish) run out in great length, lying flat in the ground, where at distances, they take root, over-spreading sometimes half a score acres, sometimes in small patches of about Rood or the like; the Leaves are like Box, but greener, thick and glistering; the Blossoms are very like the flower of our English Night Shade, after which suceed the berries, hanging by long small foot stalks. No bigger than a hair, at first they are of a pale yellow colour, afterwards red, and as big as a cherry; some perfectly round, others Oval, all of them hollow, of a sower astringent taste; they are ripe in August and September. They are excellent against the scurvy. They are also good to allay the fervour of hot Diseases. The Indians and the English use them much, boyling them with Sugar for Sauce to eat with their meat; and it is a delicate sauce, especially for roasted mutton; some make tarts with them as with Gooseberries." Cranberries were also known as Crane Berries as cranes, living in the New England bogs, ate the berries.

PIPSISSIWA

Chimaphila umbellata Pyrolaceae

The Indians gave us the common name of this plant, pipsisswa, meaning "to break up stones," which denoted its medicinal use by them as being able to "break" the stones of the bladder. A reddish color tea, of a bitter taste, was drunk by the Indians and early settlers.

Pipsissiwa

WINTER GREEN (or Checkerberry)

Gaultheria procumbens Ericaceae

An American ground cover known for its pleasing aromatic flavor and the oil of wintergreen whose best known use was as a flavoring to cover up the taste of disagreeable medicine. It has many common names, among them Creeping Wintergreen, Partridge Berry, Mountain Tea, Ground Tea and Spice Berry. The Indians searched out the biggest red berries, to eat during the winter and both the Indians and early colonists brewed a tea from the leaves.

Shrubs

ALDER

Alnus serrulata **Betulaceae**

Although William Wood wrote that "The water sprungie Alder good for nought," the Indians had found uses for this shrub. Baskets were woven of it; as a dye, the inner bark, with the addition of bloodroot, wild plum, and red osier dogwood, made porcupine quills red or yellow. Medicinally, "An Indian bruising and Cutting of his Knee with a fall, used no other remedy, than alder bark, chewed fasting and laid to it, which did soon heal it. And for Burning and Scalding, they (the Indians) first take out the fire with a strong decoction of alder bark, they then lay upon it a playster of the bark of Board Pine first boyled tender, and beat to a playster betwixt two stones." – John Josselyn.

Alder

ARROW WOOD

Viburnum dentatum Caprifoliaceae

This tree-like shrub derived its name from the fact that the Indians cut their arrows from its long, slender, straight branches.

BARBERRY (Common)

Berberis vulgaris Berberidaceae

Barberry is a European, thorn shrub, with clusters of bright scarlet berries in the fall, which became naturalized in New England. The first legislation was passed in 1670 to root it out, as it was found that wheat grown near the shrub did not thrive. The controversy raged throughout three centuries and still barberry prospered. Medicinally, it came under the Doctrine of Signature as the bright yellow wood led one to believe that it could cure the yellow disease of jaundice. The yellow wood also achieved a permanent yellow dye without benefit of a mordant. Jelly and pickles were cooked from the tart berries.

BAYBERRY

Myrica pennsylvanica Myricaceae

The bayberry, also known as Candelberry, Waxberry, or Tallowberry is best known for its early use of the wax-like berries in the making of aromatic candles. The wax for these candles was skimmed off the top of a pot of boiling water, the water remaining made a fine blue dye for woolen cloth.

BUTTON BUSH

Cephlanthus occidentalis Rubiaceae

A shrub which the bees love.

CHOKE CHERRY

Prunus virginiana Rosaceae
Wild Black Cherry – *P. serotina*

William Wood wrote: "the cherrie trees yield great store of cherries, which grow on clusters like grapes; they be much smaller than our English cherries, nothing neare so good if they be not ripe; they so furre the mouth that the tongue will cleave to the roofe, and the throate wax horse with swallowing those red bullies (as I may call them) being little better in taste. English ordering may bring them to be English cherries, but yet they are as wilde as the Indians." The natives pounded the fruits and stones to a pulp and used this paste, dried, as an addition to their pemmican; soaked and sweetened they were an addition to their winter diet. The inner bark of choke cherry made a red dye.

CRANBERRY (High Bush or Cramp Bark)

Viburnum opulus Caprifoliaceae

As the berries of the cramp bark closely resembled cranberries, the early house holders cooked fine juice, jelly and sauce from the berries.

DOGWOOD (Flowering)

Cornus florida Cornaceae

Strictly an American plant the flowering dogwood was utilized in various ways by the natives. As a medicine, the bark was a cathartic and a tonic. Because of its astringent qualities the Indians found that the split stems of the hard wood made a good tooth brush. The flowering period was a sign that frost would be unlikely and it was safe for the Indians to plant their corn.

ELDER

Sambucus canadensis Caprifoliaceae

The shrub elder is native to England and America. Indian uses included the following: "Their bowes they make of a handsome shape, strung commonly with the sinewes of Mooses: their arrowes are made of young Elderne, feathered with feathers of Eagle wings and tails, headed with brasse in shape of a heart or triangle, fastened in a slender piece of wood sixe or eight inches long, which is framed to put loose in the pithie Elderns, that is bound fast for riving; their arrows be made in the manner because the arrow might shake from his head and be left behind for their finding, and the pike only remains to gaule the wounded beast." Medicinally, the inner bark, in the form of a poultice, was applied to sore teeth. As a dye herb, depending on the mordant, the berries became blue, purple, dark rose or light lavender. The first settlers and the Indians brewed a tea of the leaves, or flowers; ate the uncooked berries; savored a jelly from the berries.

LEATHERWOOD

Dirca palustris Thymelaeaceae

The bark of this shrub, being very tough and strong, was made into thongs by the Indians.

Leatherwood

MULBERRY

Morus rubra Moraceae

The early settlers found the native mulberry delicious eating; the Indians also enjoyed the fruit.

PLUM

Beach – *Prunus maritima* Rosaceae
Wild Yellow – *P. americana*

"Plumb Tree, several kinds, bearing some long, round, yellow, red and black plums; all differing in their fruit from those in England." – John Josselyn. Esteemed as a food by the English, Edward Winslow wrote from Plymouth in 1621: "Here are grapes, white and red, and very sweet and strong Also; strawberries, gooseberries, raspis, etc.; plums of three sorts, white, black and red, being almost as good as a damson." And William Wood adds: "The Plummes of the countrey be better for Plummes than the Cherries be for Cherries, they be blacke and yellow about the bignesse of a Damson of a remarkable good taste." The Indians also enjoyed plums raw or cooked and they were often an ingredient of pemmican. From the yellow plum a red dye was derived.

POKEWEED

Phytolacca americana Phytolaccaceae

The Indians availed themselves of pokeweed for dyeing purposes. The early settlers found it to be a succulent pot herb.

ROSES

Meadow – *Rosa blanda* **Rosaceae**
Pasture – *R. humulis*
Swamp – *R. carolina*
Wild – *R. virginiana*

The Indians ate the rose hips. John Josselyn wrote of finding "Wild Damask Roses, single, but very large and sweet, but stiptick." The rose had many uses and the colonists would have made use of the wild roses of the area until the slips, brought over on succeeding ships, came to a good growth. Of these John Josselyn tells us that "English Roses (thrive) very pleasantly." A rose water sauce was added to puddings, drinks, and wine; the blossoms were candied. A Doctor's Chest of 1618 listed syrup, honey, water, vinegar and oil of roses. These were remedies for headache, insomnia, "weak and fainting spirits," a "purge for children," colds and coughs.

SARSAPARILLA

Aralia nudicaulis Araliaceae

"Sarsaparilia, a plant not yet sufficiently known by the English: Some say it is a kind of Bind Weed; we have in New England two plants (the first one Aralia hispica) that go under the name of Sarsaparilia; the one not above a foot in height without Thorns, the other having the same leaf, but is a shrub as high as a GooseBerry Bush, and full of sharp Thorns; this I esteem as the right, by the shape and flavour of the roots, but rather by the effects answerable to that we have from the other parts of the World; it groweth upon dry sandy banks by the Seaside; and upon banks of river, so far as the Salt water flowed; and within Land up in the country, as some have reported." – John Josselyn. The Indians found that eating the roots sustained them during time of war and long hunting expeditions. The early settlers learned from the Indians the value of the aromatic beverage made from the leaves, flowers, bark and roots as a cooling medicine for the blood.

Sarsaparilla

SHADBUSH

Amelanchier canadensis — Rosaceae

Also known as June-berry, Service-berry, and May-cherry, the reddish purple fruits were eaten raw by the Indians and a great many of them were dried and stored for winter use. The early settlers also ate them raw or cooked them in pastries and muffins.

Shadbush

SPICE BUSH

Lindera benzoin Lauraceae

Also called Fever Bush, the Indians valued the twigs as a relief from fevers.

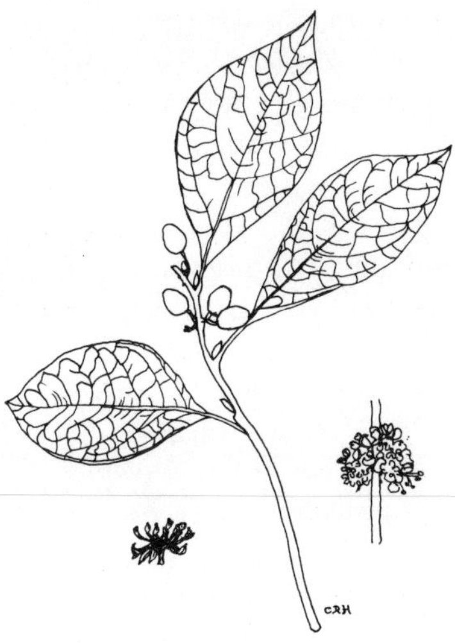

Spice Bush

STEEPLE BUSH (or Hardhack)

Spiraea tomentosa Rosaceae

Valued as a medicine since the early days as an astringent and a pain reliever.

SUMAC

Smooth – *Rhus glabra* Anacardiaceae
Staghorn – *R. Typhina*

Native both to England and New England, William Wood notes: "The Diars Shumach, with more trees there be, that are both good to use, and rare to see." John Josselyn adds: "Sumach, differing from all that I did ever see in the Herbalists; Our English Cattle devour it most abominabley, leaving neither leaf nor branch, yet it sprouts again next Spring: The English use to boyl it in Beer, and drink it for colds; and so do the Indians, from whom the English had the medicine." As a dye, the fruit produced crimson; the branches and the fruit, black; the inner bark and pulp of the stalk, a yellow which colored rushes for matting and porcupine quills. The early colonists found: "Also here are store of sumacke trees, they are good for dyeing and tanning of leather, likewise such trees yield a precious gum called wine Benjamin, that they say is excellent for perfumes." The root, also utilized as a dye, "dyeth wool cloth reddish." As a tobacco, the Indians employed a blend of dried leaves among which sumac was the most important ingredient. The Indians and early settlers enjoyed a refreshing drink which was made from the fruit. Medicinally, a liquid boiled from the fruit was beneficial to the Indians to check bleeding; leaves mixed with honey, and applied, soon healed wounds.

SWEET FERN

Comptonia peregrina Myricaceae

Although called sweet fern, this sweetly aromatic plant is not a fern but a shrub. John Josselyn gives us the following description: "Sweet Fern, the roots run one within another like a net, being very long and spreading abroad under the upper crust of the Earth, sweet in taste (the nutlets), but withal astringent, much hunted by our Swine; the people boyl the tender tops in molasses beer, and in Possets for Fluxes, for which it is excellent." To the Indians sweet fern was a cure for poison ivy; the leaves were boiled, the juice or a poultice was applied to the affected areas.

Sweet Fern

SWEET GALE

Myrica gale Myricaceae

The young buds gave the Indians a dye for porcupine quills.

SWEET PEPPERBUSH

Clethra alnifolia Clethraceae

Also known as Spiked Alder and White Alder, this shrub was named pepperbush for the resemblance of its fruits to peppercorns. The spicy, highly fragrant white flower spikes made it one of the important honey plants.

TREES

APPLE (Crab)

Pyrus coronaria Rosaceae

Although of a sour taste, jelly and tarts were cooked from the wild apple.

ASH

Black – *Fraxinus nigra* Oleaceae
White – *F. americana*

The ash, being strong and supple was employed by the Indians in the following ways: For bows and paddles; their splints, separated, made fish baskets; frames for snow shoes; pipe stems and sheets of ash were laid over the upper mats of a lodge to provide a waterproof shelter.

ASPEN (Quaking)

Populus tremuloides Salicaceae

Employed by the early settlers, under the Doctrine of Signature, as its quaking leaves led one to the belief that it would cure palsy. The Indians, in the spring, ate the sweet inner bark.

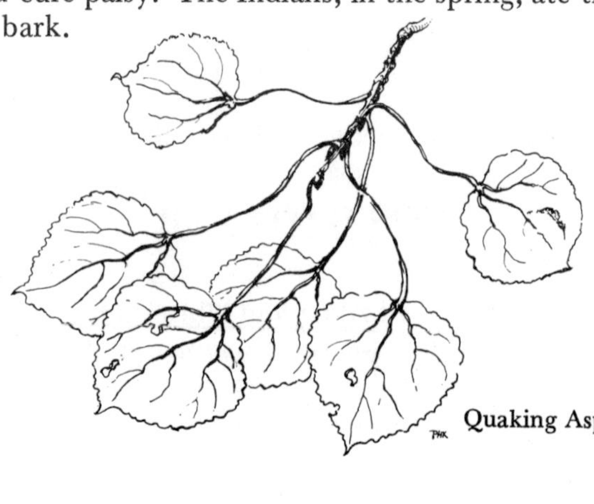

Quaking Aspen

BALSAM FIR

Abies balsamea Pinaceae

Balsam Fir yielded turpentine, pitch and tar, also fine wood to be applied in the building of ships and houses.

BEECH (American)

Fagus grandifolia Fagaceae

Indians utilized the beech as a shelter during thunderstorms as it was believed that this tree would never be struck by lightning. Bows were contrived of beech, and the tough wood was bent into shape by the Indians to make a crude type of snowshoe.

BIRCH

Gray – *Betula populifolia* Betulaceae
River – *B. nigra*
Sweet or Black – *B. lenta*
White – *B. papyrifera*
Yellow – *B. lutea*

Birch had a variety of uses to the natives. "Their other canows be made of thinne birch rines, close ribbed on the in-side with broad thinne hoopes, like the hoopes of a tub." Birch bark was made into dishes, baskets, and boxes which were used to carry water and for food and other storage. From the inner bark of white birch a dye was derived to color porcupine quills red. A tea was brewed from sweet birch twigs. Medicinally "birch, white and black; the bark of birch is used by the Indians for bruised Wounds and Cuts, boyled very tender, and stampt betwixt two stones to a plaister, and decoctions thereof poured into the Wound; and also to fetch the Fire out of Burns and Scalds. Spunk, an excrescence growing out of black birch, the Indians use it for Touch-wood; and therewith they help the Sciatica, or Gout of the Hip, or any great Ache, burning the patient with it in two or three places upon the Thigh, and upon certain Veins." – John Josselyn.

BUTTERNUT (White)

Juglans cinerea Juglandaceae

The nuts were an important food to the Indians. The hairy husks gave a good brown dye, and oil from the nut was rubbed on their hair.

CEDAR (Red)

Juniperus virginiana Pinaceae

The English were partial to red cedar as a fuel as the burning wood was very fragrant. This tree was also applied "for seeling houses, and making of chests, boxes and staves." From the bark and roots the Indians contrived rope, stamp baskets, bags and woven mats. The inner bark colored their mats red.

CHESTNUT

Castanea dentata Betulaceae

The Indians included chestnuts in their corn bread, or roasted, then ground them for a drink. Husked, dried and powdered, they were employed to thicken soups. "Chestnuts very sweet in taste and may be (as they usually are) eaten raw; the Indians sell them to the English for twelve pence the bushel." – John Josselyn. Indian boats or canoes were often constructed of chestnut which were made hollow by being burned out. They were scraped, smoothed, and shaped with sharp tools. The boats or canoes often held as many as twenty persons, were forty or fifty feet long, and as broad as the tree.

COTTONWOOD (or Cotton Tree)

Populus deltoides Salicaceae

Indian dwellings were erected of this wood.

ELM

Ulmus americana Ulmaceae
Slipper or Red Elm – *U. rubra*

Elm was considered a very fine wood for the building of houses and boats by the English; the Indians constructed their dwellings of the elm. The early settlers made chopping bowls of elm, which were scrubbed by the housewife and soon became white and shiny. Much of our knowledge of slippery elm came to us from the Indians who found it a soothing remedy and chewed the inner bark, or cooked it in fat.

HAZEL

Corylus americana Betulaceae

As a native dye, the inner bark was used with butternut to dye rushes black. The Indians esteemed the hazel nuts as a food. Medicinally the nuts were a panacea: "For sore mouths, falling of the pallat. Filberd, both with hairy husks upon the nuts, and setting hollow from the nut, and filled with a kind of water of an astringent taste; it is very good for sore mouths, and falling of the Pallat, as is the whole green nut before it comes to Kernel, burnt and pulverized. The Kernels are seldome without maggots in them." – John Josselyn.

HEMLOCK

Tsuga canadensis Pinaceae

Indian dwellings were built of hemlock. As a dye, the inner bark, with a mordant of rock dust, colored matting red. Medicinally, both the Indians and earliest settlers found that a tea brewed of the leaves was an excellent remedy for scurvy. For a burn the Indian women "playstered it with a bark of Hemlock tree, boyled soft and stampt betwixt stones, till it was as thin as a brown paper, and of the same color, she annointed the playster with Soles Oyl, and the Sore likewise," and, further, "the Indians break and heal their Swellings and Sores with it (hemlock), boyling the inner bark of young hemlock very well, then soaking it in Soyl oyl, they apply it to the Sore: It will break a Sore swelling speedily."

HICKORY (Shagbark or Shellbark)

Carya ovata Juglandaceae

The natives beat the whole hickory nuts with stones, then placed all in a mortar, added water, and ground the mixture until it became a milky substance. This was called pochicora, from which the name hickory was derived. Indian bows were carved from the second growth of white hickory. Splints of the tree were plaited into baskets.

HORNBEAM

Carpinus caroliniana Betulaceae

William Wood wrote: "The Horne-bound tree is a tough kind of wood, that requires so much paines in riving as is almost incredible, being the best for to make bottles and dishes, not being subject to cracke or leake."

JUNIPER (Common)

Juniperus communis Pinaceae

John Josselyn notes: "Juniper, which Cardanus saith is cedar in hot Countries, and Juniper in cold Countries. It is here very dwarfish and shrubby growing for the most part by the Sea Side." The English found that juniper was invaluable for fires, "keeping" they claimed "a fire a whole year without supply." The Indians never utilized the wood for this purpose.

LARCH

Larix laricina Pinaceae

Known also as Tamarack or Hackmatack. From John Josselyn: "The turpentine that issueth from the cones of the Larch tree (which comes nearest of any to the right Turpentine) is singularly good to heal wounds, and to draw out the malice of any ache rubbing the place therewith, and throwing upon it the powder of sage leaves." Indians sewed strips of the root to hold the birch bark of their canoes in place. When the early settlers started building boats, they sought out the root with an angle or bend for use in "instep Crooks" or ship's knees.

Larch

LINDEN (or Line)

Tilia americana Tiliaceae

The Indians, from the bark of Linden, made thongs, rope, string and thread. The last named sewed together the woven mats of Cattail. The bark was also useful in making Indian baskets, bags and mats. Due to the bark's mucilaginous quality, the Indians valued it as a remedy for wounds.

MAPLES

Red Maple – *Acer rubrum* Aceraceae
Sugar Maple – *A. saccharum*

Long before the early settlers arrived the Indians had found that the Maple yielded a sugar syrup. This syrup was much used as a sweetening and included a sauce to accompany their meat. The first householders learned the art of sugaring from the Indians. As a dye the Indians found the bark produced black: the rotten wood made a purple. Bowls of Maple were carved by the Indians and bartered to the settlers.

OAKS

Black Oak – *Quercus velutina* Fagaceae
Chestnut Oak – *Q. prinus*
Dwarf Chinquapin Oak – *Q. prenoides*
Post or Iron Oak – *Q. stellata*
Red Oak – *Q. rubra*
Scarlet Oak – *Q. coccinea*
Scrub or Bear Oak – *Q. ilicifolia*
Swamp White Oak – *Q. bicolor*
White Oak – *Q. alba*

"There is Oak of three kinds, white, red and black, the white is excellent to make canoes of, shallopes, ships, and other vessels for the sea, and for clawboard, and pipe staves, the black is good to make Wyanscot of: and out of the

White Oak Acorns, (which is the acorn bears delight to feed upon): The natives draw an oyl of the Moss that grows at the root of the White Oak the Indians make a strong decoction, with which they help their Papouses or young children's scall'd Heads." – John Josselyn. The Indians kept great stores of acorns for winter use. Powdered, they were added as a thickening for soup. Acorns were also employed in the preparation of bread by the Indians and the first settlers.

PINES

Pitch Pine – *Pinus rigida* Pinaceae
White Pine – *P. strobus*

"Their cannows be made either of Pine-trees, which before they were acquainted with English Tooles, they burned hollow, scraping them smooth with Clam-shels, and oyster shels, cutting their out sides with stone hatchets. These boates be not above a foot and a halfe, or two feete wide, and twenty foote long. The Firre and Pine bee trees that grow in many places shooting up exceeding high, especially the Pine: they doe afford good masts, good board, Rozen and Turpentine." "It yields a very sovereign Turpentine for the curing of desperate Wounds." – John Josselyn. Culinary uses, by the Indians, included either chewing the sap-filled inner bark or boiling it with pieces of bone. Cakes were also cooked from the bark. Smoked or dried, these were kept for winter use on long trips. The green cones, roasted, made a sweet. Both the settlers and Indians cut little slices of the wood and burned them for light. Of this latter use William Wood wrote: "but I cannot recommend it for singular good, because it is something fluttish, dropping a pitchie kind of substance where it stands."

SASSAFRAS

Sassafras albidum Lauraceae

From the following excerpt we find that Sassafras was the first export from New England: "From hence we went to another island, to the Northwest of this (Martha's Vineyard) and within a league or two of the main . . . sassafras trees, great plenty all the island over, a tree of high price and profit." – 1602. For "price and profit" the roots, in England, sold for three shillings apiece. "Buzzards Bay: But after our barke had taken in so much sassafras . . . and other commodities as were thought convenient." – 1602 "and divers other wholesome herbs, both for profit and pleasure, with great store of Sassafras." – 1623. A tea, called Saloop, derived from the roots of Sassafras, was considered a cure-all, and was sold throughout London in street stalls. Sassafras tea was considered both by Indians and early settlers as a superior spring tonic. To the Indian men, Sassafras was a remedy for rheumatism, and to the women, a tonic after childbirth.

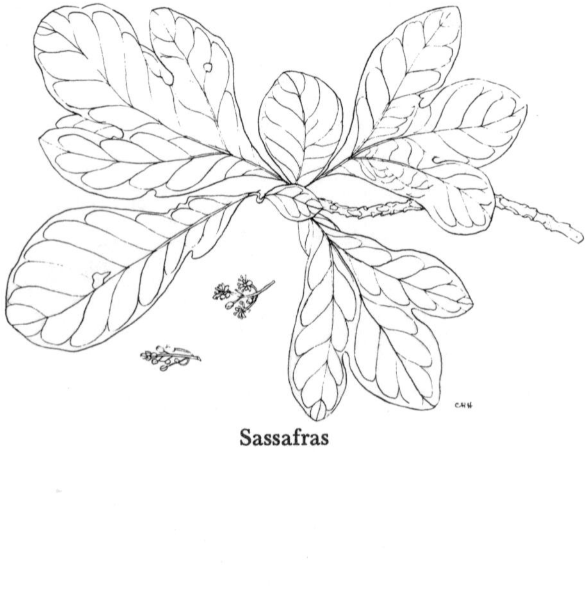

Sassafras

SPRUCES

Black Spruce – *Picea mariana* Pinaceae
Red Spruce – *P. rubens*
White Spruce – *P. glauca*

"Spruce is a goodly Tree, of which they make masts for ships, and sail Yards: It is generally conceived by those that have skill in building of ships, that here is absolutely the best Trees in the World many of them being three Fathom about, and of great length." – John Josselyn. Gum of spruce was employed by the Indians to caulk their canoes. Ropes and tackle, woven mats and thread were made from the thin spruce roots.

WALNUT (Black)

Juglans nigra Juglandaceae

"There is likewise black walnut, of precious use for tables, cabinets, and the like" and "The walnut tree is the toughest wood in the countrie, and therefore made use of for hoops." To the Indians the nuts of the Walnut were a principle wild food: one use being to thicken their soups. As William Wood wrote: "These trees have a very good Nut, something smaller, but nothing inferior in sweetness and goodnesse to the English nut, having no bitter pill." As a native dye the Indians made black from the root, and brown from the leaves, husks, and whole nut. For dwellings: "The Natives of New England are accoustomed to build their houses much like the wild Irish: they gather poles in the woodes and put the great end of them in the ground placing them in form of an Arch, then Bind them together with the Barke of walnut trees, which is wondrous tuff." – *New England Canaan, 1622.*

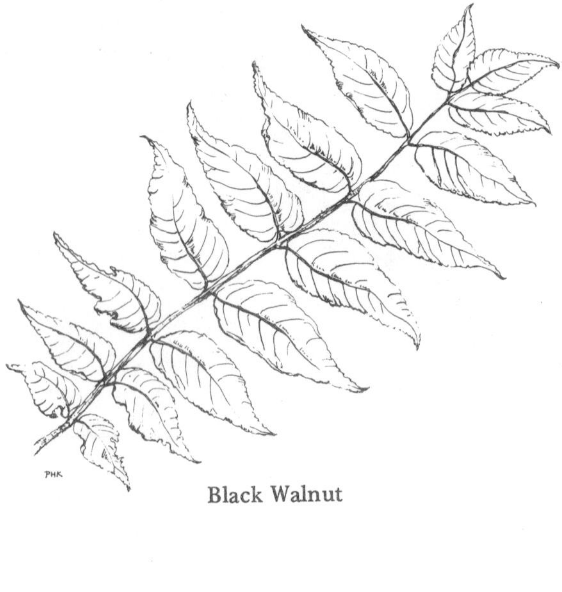

Black Walnut

WHITE THORN (Pilgrim's)

Crataegus coccinea Rosaceae

The early colonists called this native Hawthorn the Pilgrims' Whitethorn, as it reminded them of the English Hawthorn, or Mayflower (*Crataegus oxycanthus*) for which the ship Mayflower was named. They also found the haws nearly as good to eat as their English cherries.

Hawthorn

WILLOW

Salix spp. Salicaceae

The Indians made bows and snowshoes of several species of the supple willow, which also lent itself to basketry. The latter were beautified by stripping the willow during different seasons and employing native dyes. Dolls, for the Indian children, were also shaped from willow.

WITCH HAZEL

Hamamelis virginiana Hamamelidaceae

Bows were fashioned by the Indians of witch hazel. For medicine the Indians used this shrub-like tree in the form of a poultice for external irritations and bruises. The inner bark, in the same form, removed inflammation of the eyes: painful tumors were also treated. A tea of the aromatic dried leaves, sweetened with maple sugar, was drunk by the Indians. From the earliest days of the colonists a forked branch, carried over the ground, indicated water and even mineral deposits.

Witch Hazel

YEW (American or Canadian)

Taxus canadensis Pinaceae

The Indians fashioned their bows from yew.

BIBLIOGRAPHY

ed. AMERICAN HERITAGE. *The American Heritage Cook Book*
American Heritage Publishing Co., Inc., Simon & Schuster, 1964.

BLANCHAN, Neltje. *Nature's Garden*
New York. Doubleday, Page & Company, 1900.

ed. BOTKIN, B. A. *A Treasury of New England Folklore*
Crown Publishers, New York.

BRADFORD, William. *Of Plymouth Plantation, 1620-1647*
New edition with notes and introduction by Samuel Eliot Morison.
New York, Alfred A. Knopf, 1952.

BRERETON, John. *Discovery of the North Part of Virginia.* 1602
Collections of the Massachusetts Historical Society, Volume VIII.
Boston, Charles C. Little & James Brown, 1843.

BROWN, Alice Cooke. *Early American Herb Recipes*
Charles E. Tuttle Company, Rutland, Vermont, 1966.

BROWNLOW, Margaret E. *Herbs and the Fragrant Garden*
The Herb Farm Ltd., Seal, Sevenoakes, Kentucky, 1965.

CARRIER, Lyman. *The Beginnings of Agriculture in America*
McGraw-Hill Book Company, Inc., New York.
London. 6 and 8 Briverie Street, 1923.

CHAMPLAIN. *Champlain's Voyages, 1604-6*
The publication of the Prince Society, Boston.
John Wilson & Son, 1880.

CHEEVER, George B. *The Pilgrim Fathers – of the Journal of the Pilgrims at Plymouth New England in 1620*
New York. John W. Lowell Company, 1848.

CLAIR, Colin. *Of Herbs and Spices*
Abelard-Schuman. London. New York. Toronto, 1961.

CLARKSON, Rosetta E. *Magic Gardens*
New York. The Macmillan Company, 1942.

CLUTE, Willard Nelson. *Our Ferns*
Frederick A. Stokes Company, New York, 1938.

COLES, William. *The Art of Simpling*, 1657
Reprinted by the Herb Grower Press, Falls Village, Conn., 1966.

COLLINGWOOD, G. H. and BRUSH, Warren D. *Knowing Your Trees* The American Forestry Association, Washington, D. C., 1947.

COON, Nelson. *Using Plants for Healing* Hearthside Press, Inc., 1963.

COON, Nelson. *Using Wayside Plants* Eaton Press, Inc., Watertown, Massachusetts, 1959.

CRAWFORD, Mary Caroline. *In the Days of the Pilgrim Fathers* Boston. Little Brown and Company, 1920.

CROSS, Chester. *Weeds of the Massachusetts Cranberry Bogs. Part l: The Grasses* Agricultural Experiment Station, University of Massachusetts, Amherst, Massachusetts, April, 1952.

CULPEPER, Nicholas. *Culpeper's Complete Herbal*, 1653 London W. Foulsham & Co., Ltd. Distributed by Sterling Publishing Co., Inc., 419 Fourth Avenue, New York.

CUTLER, Manasseh. *An Account of Some of the Vegetable Productions Naturally Growing in this Part of America.* Bulletin of the Lloyd Library of Botany, Pharmacy and Materea Medica. J. O. and C. G. Lloyd, Cincinnati, Ohio, 1785.

DAME, Lorin P. and BROOKS, Henry. *Handbook of the Trees of New England* Boston. Ginn and Company, The Athenaeum Press, 1902.

DANA, Mrs. William Starr. *How to Know the Wild Flowers* New York. Charles Scribner's Sons, 1895.

DODOENS, D. Remberte. *A New Herbal or History of Plants, etc.* London, 1619.

DOW, George Francis. *Everyday Life in the Massachusetts Bay Colony* Boston. The Society for the Preservation of New England Antiquities, 1935.

DRIVER, Harold E. and MASSEY, William C. *Comparative Studies of North American Indians* Transaction of the American Philosophical Society held at Philadelphia, for the promoting useful knowledge. Volume 47, part 2, 1957.

EARLE, Alice Morse. *Old Time Gardens* New York. The Macmillan Company, 1901.

EMERSON, George B. *A Report on the Trees and Shrubs Growing Naturally in the Forests of Massachusetts*
Boston. Little, Brown and Company, 1875.

EVELYN, John. *Acetaria. A Discourse of Sallets.* 17th Century
Brooklyn. Published by the Women's Auxiliary, Brooklyn Botanic Garden, 1937.

EYLER, Ellen C. *Early English Gardens and Garden Books*
The Folger Shakespeare Library, Cornell University Press, 1963.

FARWELL, Edith Foster. *My Garden Gate Is on the Latch*
Copyright, 1962. Edith Foster Farwell.

FERNALD, Merritt Lyndon and KINSEY, Alfred Charles.
Edible Wild Plants. Revised by Reed C. Rolling.
Harper & Brothers, Publishers, New York, 1958.

FLAG, Wilson. *The Woods and By-Ways of New England*
Boston. James R. Osgood and Company, 1872.

FLANNERY, Regina. *An Analysis of Coastal Algonquin Culture*
The Catholic University of America. Anthropological Series, No. 7, 1939.

FLETCHER, H. L. V. *The Fragrant Garden*
The Garden Book Club, 121 Charing Cross Road, London W.C. 2.

FLINT, Martha Bockee. *A Garden of Simples*
Charles Scribner's Sons, New York, 1900.

FOSTER, Gertrude B. *Herbs for Every Garden*
New York. E. P. Dutton and Company, Inc., 1966.

FOX, Helen M. *The Years in My Herb Garden*
New York. The Macmillan Company, 1953.

FREEMAN, Margaret B. *Herbs for the Mediaeval Household*
Metropolitan Museum of Art, New York, 1943.

GATHERCOAL, E. N. *Check List of Native and Introduced Drug Plants in the United States*
Massachusetts College of Pharmacy, Chicago, 1942.

GIBBONS, Euell. *Stalking the Healthful Herbs*
David McKay Company, Inc., New York, 1966.

GIBBONS, Euell. *Stalking the Wild Asparagus*
David McKay Company, Inc., New York, 1965.

GILMORE, Melvin R. *Indian Lore and Indian Gardens*
The Slingerland-Constack Company, Ithaca, New York, 1930.

GOODWIN, John A. *The Pilgrim Republic. An Historical Review of the Colony of New Plymouth*
Boston. Ticknor and Company. London, Trubner and Company, 1888.

GOOKIN, Daniel. *Historical Collections of the Indians in New England,* 1674
Collections of the Massachusetts Historical Society for the year 1792. Vol. 1.

GOTTSCHO, Samuel. *Wild Flowers*

GRIGSON, Geoffrey. *The Englishman's Flora*
Phoenix House, Ltd., London 1958 & 1960.

GRIEVE, Mrs. M. *Culinary Herbs and Condiments*
New York. Harcourt Brace and Company, 1934.

GRIEVE, Mrs. M. *A Modern Herbal.* Volumes 1 and 2.
Hafner Publishing Company, New York, 1959.

HALLETT, Leaman F. *Medicine and Pharmacy of the New England Indians*
Bulletin of the Mass. Archaeological Society, Vol. 17, No. 3 April, 1956.

HARTLEY, Dorothy. *Food in England*
Macdonald. London, 1964.

HEDGE, Catherine Elliott. *Wild Flowers of Plymouth & Vicinity, 1804-1904*
Boston. George H. Ellis Co., Printers, 272 Congress Street, 1904.

HEDRICK, U. P. *A History of Horticulture in America to 1860*
New York. Oxford University Press, 1950.

HIGGESON, Mr. *New England's Plantation*, 1629
Collections of the Mass. Historical Society for the Year 1792. Volume 1.

HOITT, S. W. *Native Shrubs for Northern New England*
Published and distributed by the University of New Hampshire, Durham, N. H. S. W. Hoitt, Director of the Cooperative Extension Service, the United States Department of Agriculture cooperating.

JOSSELYN, John. *New England's Rarities,* 1672.
Boston. William Veazie, 1865. With an Introduction and Notes by Edward Tuckerman.

JOSSELYN, John. *An Account of Two Voyages to New England, 1638. 1663*
Boston. William Veazie, 1865.

KELLER, Harriet L. *Our Native Trees*
Charles Scribner's Sons, New York, 1902.

KEPHART, Horace. *Camping and Woodcraft*
New York. The Macmillan Company.

KIERAN, John. *An Introduction to Wild Flowers*
Hanover House, Garden City, New York.

KIERSTEAD, Sallie Pease. *Natural Dyes*
Boston. Bruce Humphries, Inc., 1950.

KREIG, Margaret B. *Green Medicine*
Rand McNally & Company, New York, Chicago, San Francisco, 1964.

LEVETT, Christopher. *A Voyage into New England, 1623-24*
Collections of the Massachusetts Historical Society, Volume VIII. Boston. Charles C. Little & James Brown, 1843.

LEVEY, Juliette de Bairacli. *Herbal Handbook for Everyone*
Charles T. Branford Company, Newton, Massachusetts, *1966.*

LEYEL, Mrs. C. F. *Herbal Delights*
Faber and Faber, Limited, 24 Russell Square, London, 1950.

LUCAS, Jannette May. *Indian Harvest*
J. B. Lippincott Co., Philadelphia and New York, 1945.

LYFORD, Carrie A. *Ojibwas Crafts*
Department of the Interior.

MATHEWS, F. Schuyler. *Field Book of American Trees and Shrubs*
G. P. Putnam's Sons, New York and London, 1915.

MEYER, Joseph E. *The Herbalist*
Printed and Bound by Rand McNally & Company, Conkey Division, 1960.

MILORADOVICH, Milo. *Herbs and Spices*
Doubleday, 277 Park Avenue, New York, New York.

MORTON, Thomas. *The New England Canaan*, 1622
Publication of the Prince Society, Boston. John Wilson & Son, 1883.

NORBECK, O. E. *Book of Indian Life Craft*
Associated Press, New York, 1958.

PARSONS, Frances Theodora. *How to Know the Ferns*
Dover Publications, Inc., New York, New York, 1961.

PEATTIE, Donald Culross. *A Natural History of Trees*
Houghton Mifflin Co., Boston, 1950.

PLATT, Rutherford. *A Pocket Guide to Trees*
Dodd Mead & Company, 1953.

PRING, Martin. *Sailors' Narratives of Voyages Along the New England Coast 1524-1624*
With Notes by George Packer Winship.
Boston. Published by Houghton Mifflin & Company, 1905.

RICKETT, Harold William. *Wild Flowers of the United States, volume 2 - The Northeastern States*
McGraw-Hill Book Company.

RICKETT, Harold William. *The Odyssey Book of American Wildflowers.*
The Odyssey Press, New York, 1964.

ROHDE, Eleanour Sinclair. *Garden of Delight*
London. The Medici Society, Ltd. Boston. A. Cushman & Flint, 1934.

ROHDE, Eleanour Sinclair. *Garden of Herbs*
Herbert Jenkins Limited, 3 York Street, St. James, London, 1926.

ROHDE, Eleanour Sinclair. *The Scented Garden*
The Medici Society, Limited, 7 Grafton Street, London.

ROSIER, James. *The Land of Virginia*, 1605
Collections of the Massachusetts Historical Society, Vol. VII.
Boston. Charles C. Little and James Brown, 1843.

ROWSE, A. L. *The Elizabethans and America*
Harper & Brothers, Publishers, New York, 1959.

SANFORD, S. N. F. *New England Herbs*
New England Museum of Natural History, 234 Berkeley Street, Boston, 1937.

ed. SEYMOUR, E. L. D. *The Garden Encyclopedia*
New York. William H. Wise & Company, 1937.

SLADE, Daniel Denison. *The Evolution of Horticulture in New England*
New York and London. Putnam's Sons, New Rochelle, New York. The Knickerbocker Press, 1895.

STEP, Edward. *Herbs of Healing*
Hutchinson and Co., Ltd., London. New York, 1950.

TAYLOR, Kathryn S. and HAMBLIN, Stephen F. *Handbook of Wild Flower Cultivation*
The Macmillan Company, New York, 1963.

TAYLOR, Norman. *Wild Flower Gardening*
S. Van Nostrand Co., Inc., Princeton, New Jersey. Toronto, London, New York, 1955.

TAYLOR, Raymond L. *Plants of Colonial Days*
Williamsburg. Printed for Colonial Williamsburg, 1952.

TUNIS, Edwin. *Woodland Hunters*

VANDERHOOP, M. A. C. *History and Tradition of the Gay Head Indians*

WILDER, Louise Beebe. *The Fragrant Path*
New York. The Macmillan Company, 1932.

WILLIAMS, Roger. *A Key into the Language of America*
London. Printed by Gregory Dexter, 1643. Reprinted at Providence for the Rhode Island and Providence Plantations Tercentenary Committee, Inc., 1936.

WILLOUGHBY, Charles C. *Antiquities of the New England Indians*
Peabody Museum of American Archaeology and Ethnology, Harvard University, Cambridge, Mass., 1935.

WOOD, William. *New England's Prospect, 1629-34*
Reprinted Boston, Massachusetts, December 22, 1897.

WREN, R. C. *Potters New Cyclopedia of Botanical Drugs & Preparations*
Pitman Publishing Corp., New York. Toronto. London for Potter & Clarke, 1956.

Dye Plants and Dyeing
Brooklyn Botanic Garden Record. Autumn, 1964. Volume 20, number 3.

Plants and Gardens
Brooklyn Botanic Garden Record. Autumn, 1950 New Series. Volume 6, number 3.

INDEX OF COMMON & LATIN NAMES

G

H

I

J

L

M

N

O

P

NOTES

NOTES

NOTES

NOTES